MYSTERIES OF THE SATELLITES

ALSO BY FRANKLYN M. BRANLEY

MYSTERIES OF THE UNIVERSE SERIES

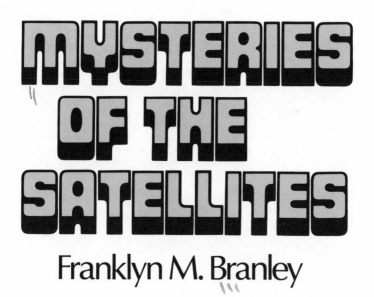

MYSTERIES OF THE SATELLITES

Franklyn M. Branley

Diagrams by Sally J. Bensusen

LODESTAR BOOKS   E. P. DUTTON   NEW YORK

Text copyright © 1986 by Franklyn M. Branley
Illustrations copyright © 1986 by E. P. Dutton

Library of Congress Cataloging in Publication Data

Branley, Franklyn Mansfield, date.
  Mysteries of the satellites.

  (Mysteries of the universe series)
  Bibliography: p.
  Includes index.
  Summary: Discusses the origins and characteristics of the natural satellites that travel around seven of the nine planets.
    1. Satellites—Juvenile literature. [1. Satellites]
I. Bensusen, Sally J., ill. II. Title. III. Series:
Branley, Franklyn Mansfield, date. Mysteries of the universe series.
QB401.5.B73 1986 523.9'8 85-20770
ISBN 0-525-67176-5

Published in the United States by E. P. Dutton,
2 Park Avenue, New York, N.Y. 10016

Published simultaneously in Canada by
Fitzhenry & Whiteside Limited, Toronto

Editor: Virginia Buckley   Designer: Riki Levinson

Printed in the U.S.A.   W   First Edition
10 9 8 7 6 5 4 3 2 1

Photograph on opposite page
courtesy of Jet Propulsion Laboratory

CONTENTS

1 THE PLANETS AND THEIR SATELLITES

Volcanoes that shoot out hot liquid sulfur, holes large enough to swallow an entire village, miles and miles of wasteland, worlds of ice and poison gases—all are found among the satellites of the planets. Some satellites are huge, thousands of miles across, and are even larger than the smaller planets. Others are only a few miles in diameter. All are dependent upon the planets. *Satellite* means dependent, a companion or follower.

How many satellites are there?

Among the nine planets, seven of them have one or more satellites. Only Mercury and Venus have none. Mercury has low gravity, probably not enough to hold a satellite. At one time, Venus may have had a satellite, but long ago it may have shattered and fallen as ash and dust onto the planet itself.

The satellites of the planets are listed on pages 6 and 7. These are natural satellites. Earth also has about a thousand man-made satellites going around it. Distributed among the planets are forty-four satellites; thirty-three of them belong to two of the planets—Jupiter has sixteen and Saturn has seventeen, although these planets may have several more. It is very likely that other small satellites will be revealed as new space probes move in close to the planets, and as the Hubble Space Telescope, planned to be launched by shuttle in 1986, takes a close look at them. Probes can only take snapshots of the planets and their vicinity. The telescope is able to watch them for hours, and so may be able to see objects that cannot be revealed by snapshots.

Many astronomers would not be surprised if Jupiter and Saturn, the two largest planets, have a good many more satellites than have already been identified.

Where did satellites come from?

No one has the final answer to this question. It is believed that many of the satellites may have formed from cosmic dust clouds. Space is not empty. It contains tremendous amounts of dust and gases. Even now, some 20 000 tons of meteoric dust falls upon Earth every year. Large amounts are also attracted to the other planets and to some of the larger satellites.

Before there were planets, cosmic dust clouds were extremely dense and extensive. Particles in the clouds were in motion. On occasion, two or more particles collided in such a way that they joined together. This made a more massive concentration in the cloud. Its gravity was great enough to attract other particles. As mass increased, so did gravity, and the concentration grew in size.

It is reasonable to suppose that the planets formed in some such fashion. Smaller amounts of material became satellites of the planets.

Perhaps the Moon was formed in this way. Yet mysteries remain, for among the satellites, there are some that appear to have had different beginnings. For example, Phobos and

Deimos, the satellites of Mars, may be the remains of what was once a single satellite. Both are small and, more important, both are very irregular. Some people refer to them as potato satellites because of their shapes and uneven surfaces. Perhaps a larger satellite exploded or was ripped apart by the gravity of Mars. Or Mars may have collided with a larger object that has since gone into a distant orbit, and Phobos and Deimos may be debris remaining after the collision.

Some of the satellites of the outer planets, such as Pluto's satellite, Charon, may have had other origins. Far out beyond the solar system and surrounding it, there is a great cloud of dust particles. It is believed that this cloud is the breeding ground of comets. There may be sufficient dust in the cloud to produce billions of comets. Perhaps, long ago, Pluto captured a cometlike mass, or several that joined together, and that settled into an orbit circling Pluto and became its satellite. Pluto itself may have escaped from Neptune's orbit; maybe Charon did also.

In the region between the orbits of Mars and Jupiter, there are untold numbers of asteroids. These are masses of stone and metal which may be up to 800 kilometers in diameter, although most are smaller. They are called asteroids because, in a telescope, they appear starlike—*aster* means star and *oid* means resembling. For the most part, asteroids are in orbit around the Sun. Occasionally, a planet may pull an asteroid out of its solar orbit. This maverick asteroid may crash into the planet, or it may go into orbit around the planet.

There are at least forty-four satellites in the solar system. No doubt they have had different histories. We'll look more closely at some of them and explore some of their mysteries.

But, just as detectives cannot solve all mysteries, we'll not be able to answer all questions about the satellites. To solve a crime, detectives gather all the information they can. In a similar way, scientists try to collect and interpret information. Just as

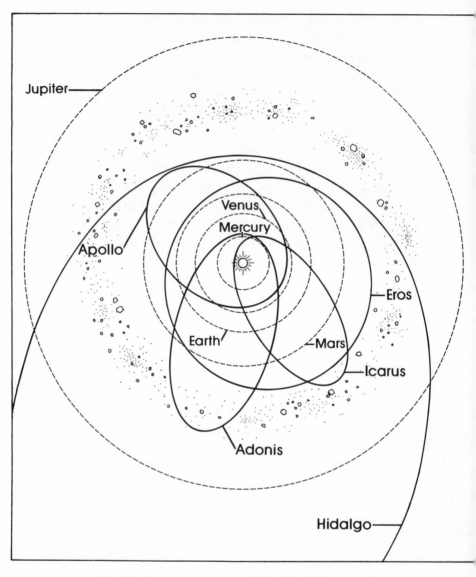

The dashed lines show the orbits of Mercury, Venus, Earth, Mars, and Jupiter. The asteroids are clustered in the region between the orbits of Mars and Jupiter. The solid lines show occasional asteroids that wander through other regions of the solar system.

there are many ways a detective may build a case around the gathered information, there are often many ways to explain mysteries of science. The correct solution is not always apparent.

Presently, at least forty-four satellites are distributed among seven of the planets. Mercury and Venus have no satellites.

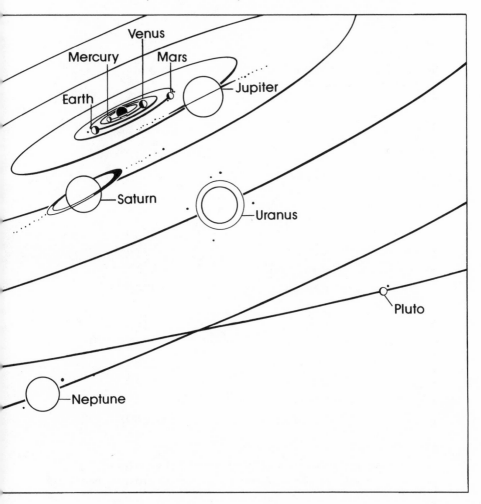

NATURAL SATELLITES
OF THE PLANETS*

	DIAMETER KILOMETERS
Earth (1)	
Moon	3476
Mars (2)	
Phobos	27
Deimos	16
Jupiter (16)	
Metis	40
Adrastea	24
Amalthea	268
Thebe	109
Io	3630
Europa	3138
Ganymede	5262
Callisto	4800
Leda	9
Himalia	180
Lysithea	20
Elara	80
Ananke	18
Carme	30
Pasiphae	40
Sinope	30

*As new instruments develop, additional satellites may be discovered. The above satellites are listed in order of increasing distance from their planets.

	DIAMETER KILOMETERS
Saturn (17)	
Atlas	38
1980 S 27	140
1980 S 26	110
Janus	218
Epimetheus	136
Mimas	392
Enceladus	500
Tethys	1060
Telesto	25
Calypso	38
Dione	1120
1980 S 6	36
Rhea	1530
Titan	5150
Hyperion	350
Iapetus	1460
Phoebe	218
Uranus (5)*	
Miranda	500
Ariel	1290
Umbriel	1210
Titania	1600
Oberon	1630
Neptune (2)	
Triton	3500
Nereid	400
Pluto (1)	
Charon	1200

*Photographs from Voyager 2's 1986 encounter with Uranus indicate that ten more satellites are circling the planet.

 # THE STORY
OF THE MOON

How old is the Moon?

Since the oldest rocks found on Earth are 3.8 billion years old, the planet must be at least that old. Moon rocks are older. The oldest rocks found there are 4.5 billion years old.

Most of Earth's original rocks have been worn away by wind and water, or they are now deep inside the planet. So there may be rocks much older than those that have been found, but they are no longer on those parts of Earth we can reach. Rocks on the Moon have not been displaced, and there has been very little erosion, because the Moon lacks wind and water. Old rocks are still on the surface. It is very likely that Earth and the Moon are the same age, that they were formed at the same time. But the two have had different histories, as you can see by looking at the craters of the Moon and the seas and continents of Earth.

Where did the Moon come from?

The origin of the Moon is a mystery. However, there are scientists who believe it used to be part of Earth. They figured out the vol-

On the Moon, sunlight is intense; shadows are black and cold. Here, Astronaut Harrison H. Schmitt explores the region around a huge, split lunar boulder—which may be a fragment of an asteroid. The Lunar Rover is at the right. NASA

ume of the Moon and found that the satellite would fit into the basin of the Pacific Ocean. During the stresses and strains of the early Earth, the crust could have been weak. A massive object such as a star may have passed close to the Sun and exerted a strong pull on Earth. The pull may have been great enough to break off a large piece of the planet, a mass that then became the Moon. Or during its early history, Earth may have been spinning much faster than it does now. It may have spun so fast that it broke apart, and a smaller part of it may have become the Moon.

If either of these two events happened, we would expect to find that the minerals on the Moon are the same as those found on Earth. But they are not. The same elements are there, such as oxygen, silicon, iron, and aluminum. But the relative amounts are different. There is also no water in the crystals found in Moon rocks, but there is water in Earth rocks.

A good many people believe that Earth and the Moon, the Sun, and all the planets originated at about the same time and that they all came from the same basic materials. Some 5 billion years ago, they say, all the objects in the solar system formed from a mass of gases, mostly hydrogen and helium, that was 15 billion kilometers across. Many of the gases escaped, and the remainder shrank to the present size of the solar system, which is about 12 billion kilometers across. As the gases contracted, they spun faster and faster. They collapsed and flattened into a disk. The larger central mass of gases packed together and became the Sun. Smaller concentrations became the planets, and much smaller amounts became satellites of the planets.

In those planets close to the Sun—Mercury, Venus, Earth, and Mars—the materials packed close together, giving them high densities. Materials are spread out more thinly in the outer planets; they have low densities. For example, the density of Earth is 5.5, meaning that its material is packed 5.5 times closer together than is the matter in water. The density of Saturn is 0.69, less than that of water. The density of the Moon is 3.36; it is made of material that is less massive than that which makes up Earth.

The Moon could be made of material that was left over after Earth had formed, or it could be made of rocks from the upper layers of Earth. That's the belief of the collision theory. It says that in the early days of Earth, after most of the iron and nickel had sunk to the core and before the Moon had formed, a large massive object crashed into Earth. The object may have been half the size of Earth—as big as Mars. The two objects did not collide head-on; more likely they grazed each other. But the impact was great enough to vaporize part of the mass, as well as part of Earth itself. The gigantic explosion threw much of the material out great distances and into an Earth-circling orbit.

The Moon contains a greater percentage of rocky material than does Earth. According to the collision theory, such a difference would be expected. In the colliding mass, as well as in

The Moon may be debris that was produced when a huge mass, perhaps as large as Mars, dealt Earth a glancing blow and caused great heat and a tremendous explosion.

Earth, much of the iron would have already gone to the central core. Rocky material in the outer layers, rather than heavy interior metals, would be involved in the explosion caused by the collision.

Similar collisions may have also produced some of the satellites of other planets. Such a theory could help explain differences among the satellites. A theory of the same origin for all planets and satellites would lead one to expect greater similarities than do exist.

Perhaps there was a tremendous collision long ago, and perhaps not. The mystery of the beginning of the Moon remains with us.

Presently, the Moon appears to be entirely solid. Whatever water or atmosphere it may have had is gone. The Moon's gravitation is not strong enough to keep them from escaping. Gases and water may have emerged later from the interior of the Moon, but if so, they have also been lost. Throughout most of its existence, if not forever, the Moon has been a dead, lifeless world—one without an atmosphere and without water.

These are theories about the Moon, suggestions based upon our present knowledge. They may be right. Or there may be other explanations about why the Moon exists and what has happened to it through the years. It is very likely that other Moon landings will be made before the end of this century, and explorers may be able to piece together more parts of the puzzle.

How has the Moon changed?

If the Moon and Earth had a similar origin, why are the two so different? At one time, hydrogen, helium, and probably gaseous water surrounded the Moon. But the Moon could not hold them. Air and water are the two principal causes of erosion. They wear down rocks and hills, and move sand and silt from one place to another. Erosion causes the high parts of Earth to be worn down and the low parts to be filled in. Carried to its extreme, erosion would smooth off the surface of our planet.

Impact craters are prominent on the Moon. They were gouged out when great masses of stone and metal, perhaps stray asteroids, crashed into the Moon millions of years ago. In a similar fashion, Earth was also bombarded. Because there is very little erosion on the Moon, today the craters appear very much as they did when they were formed.

Most of the craters that were gouged out of Earth are no longer here. They have been worn away or filled in with sand and silt. Around the world, only seventy craters are preserved well enough to be identified. All of them are in rocky locations. Twenty of them are in the hard granite that stretches across much of Canada. The most famous crater in the United States is at Canyon Diablo near Winslow, Arizona.

The surfaces of Mercury and Mars also reveal heavy cratering, as do the satellites of Mars, Jupiter, and Saturn. So far as we have seen, all objects in the solar system have been bombarded. Wherever erosion has been slight, the craters remain much as they have been since they were first formed.

In addition to impact craters, the surface of the Moon also contains volcanic craters. At one stage in its history, the Moon must have had many volcanoes that threw rocks and ash over the landscape. The ash spread far and wide and tended to level off those places where it fell. This may have been how some of the seas, or mares, were formed. They are called seas because of their relative smoothness, not because they contain water. The names of some of the seas and craters of the Moon are shown in the photos on pages 14 and 15.

Notice that the features on the right side bear pleasant names, such as Sea of Nectar, Sea of Serenity, Sea of Tranquility. Those on the left side have more ominous names, such as Sea of Rain, Ocean of Storms, Sea of Clouds. The right-hand features appear during first quarter Moon, while those on the left are seen during last quarter. According to ancient beliefs, the first quarter Moon brought good times and good weather, for the Moon was reappearing. At the time of last quarter, there was alarm, because the Moon was disappearing. Weather supposedly became poor, there were clouds and rain, and all conditions were dismal. Maybe so. It is up to you to decide whether the beliefs were true or not.

Plato

Sea of
Rains

Archimedes

Kepler

Copernicus

Ocean of
Storms

Ptolemy

Sea of
Moisture

Sea of
Clouds

The Moon in the last quarter, age twenty-two days   LICK OB-

14   *SERVATORY*

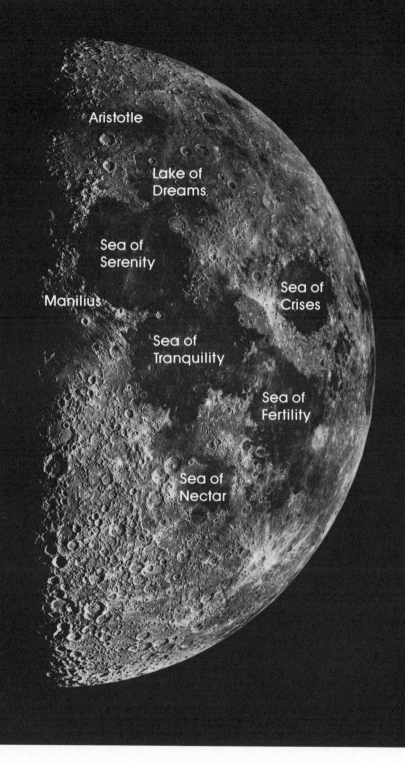

Aristotle

Lake of
Dreams

Sea of
Serenity

Manilius

Sea of
Crises

Sea of
Tranquility

Sea of
Fertility

Sea of
Nectar

The Moon in the first quarter, age seven days   *LICK OBSERVATORY*

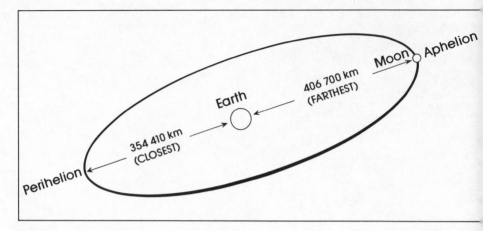

The distance between Earth and the Moon varies. The average distance is 384 320 kilometers.

What are the Moon's motions?

Earth's gravity holds the Moon in orbit. The path it follows is an ellipse, causing the distance between Earth and the Moon to vary. Each month, there is a close approach, called perigee, and a distant position, called apogee. The difference between the two is about 52 000 kilometers.

The average distance to the Moon is 384 320 kilometers. But this was not always so. It is very likely that about a billion years ago, the Moon was only 19 000 kilometers away. It would have been an immense object in our sky, twenty times larger than it appears today. Its pull on Earth would have been tremendous, and Earth's pull on the Moon also would have been very great.

At one time, the Moon was probably much closer to us. If so, ▶ *it would have appeared very large in our sky—perhaps twenty times larger than it now appears.*

16

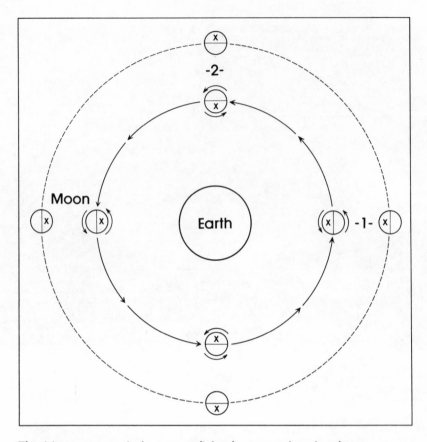

The Moon rotates (spins around) in the same time it takes to revolve (go around Earth). Therefore, the same face of the Moon is always turned toward us.

In moving from 1 to 2, the Moon revolves one-fourth of the distance around Earth, and it rotates one-fourth of the way around itself.

If the Moon rotated faster, the face marked X would alternately be toward and away from Earth.

At its present location, Earth's attraction causes the Moon to rotate once in 27.3 days, which is the same time it takes the Moon to go once around Earth. Because the periods of rotation and revolution are the same, the same face of the Moon is always turned toward us. From Earth, we cannot see the far side of the Moon, although it has been photographed. The picture on this page shows the far side. It is more heavily cratered than the side that is familiar to us.

The far side of the Moon has a rugged surface made of craters superimposed on older craters. This photograph, taken from Apollo 8, shows the lunar far-side terrain illuminated by the Sun directly overhead. NASA

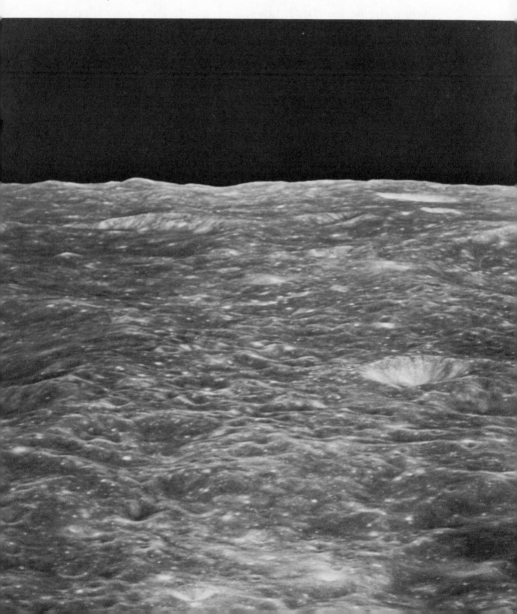

The constant pull of Earth on the Moon has caused the Moon to be somewhat deformed. It is not a sphere, for the Moon has a great lump, or bulge, in line with Earth. The steady pull of Earth has caused the Moon to bulge along its equator.

At the same time, the Moon's gravity (as well as that of the Sun) causes tides on Earth's surface. It also causes tides in the oceans. They vary a great deal during a month because of the changing position of the Moon and changes in the Earth-Moon distance. When the Moon is close to us, at perigee, and when the Moon is full, tides are highest. They are called perigee high tides.

The Moon moves west to east around Earth. However, the most apparent motion to you and me is in the opposite direction, from east to west. This motion is apparent; it only seems to happen because Earth is rotating. In other words, moonrise and moonset result from the turning of our planet.

You can check the real west-to-east movement of the Moon quite easily. The next time you see the Moon, make a sketch

The apparent motion of the Moon from east to west results from Earth's rotation (dashed line). The real motion of the Moon around Earth is from west to east (dotted line). The Moon moves about 13 degrees a day, 360 degrees a month.

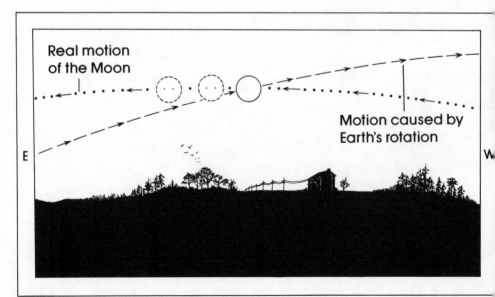

Real motion of the Moon

Motion caused by Earth's rotation

E   W

showing its location among nearby stars. The next night, mark the Moon's location on your sketch. You will see that it has moved eastward. It takes 27.3 days to go once around us, to travel through 360 degrees. Therefore, in one day the Moon moves 360 divided by 27.3, or about 13 degrees. (The distance between the pointer stars of the Big Dipper is 5 degrees, so in twenty-four hours the Moon moves about two and a half times that distance.) It's quite a lot, as you can readily see from your sketch of the sky and the Moon's changing locations.

Why does the apparent shape of the Moon seem to change?

The west-to-east motion of the Moon is related to the Moon's phasing, the changes in its appearance as the nights go by. These changes used to be a major mystery. At one time, people believed the Moon was a god. As the Moon became a slim crescent, they thought the god was leaving the sky. There was a time of dread. Once the new crescent appeared, all was well.

Today we understand why the phases occur. A cycle begins when the slim crescent Moon appears in the western sky shortly after sunset. You may be able to see it while the Sun is still above the horizon, but it will be more apparent after the Sun has set and the sky has darkened.

About a week later, the Moon will have moved one-fourth of the way around Earth. At sunset, it will be visible toward the south, and it will appear as a quarter Moon.

Another week later, the Moon will be halfway along its orbit. It will rise in the east as the Sun sets in the west. This is the full Moon.

And yet a week later, the Moon will be three-fourths of the way around its orbit. We will see it as a quarter Moon rising around midnight. It is the third or last quarter.

After about four weeks (27.3 days), the Moon has completed its orbit around Earth.

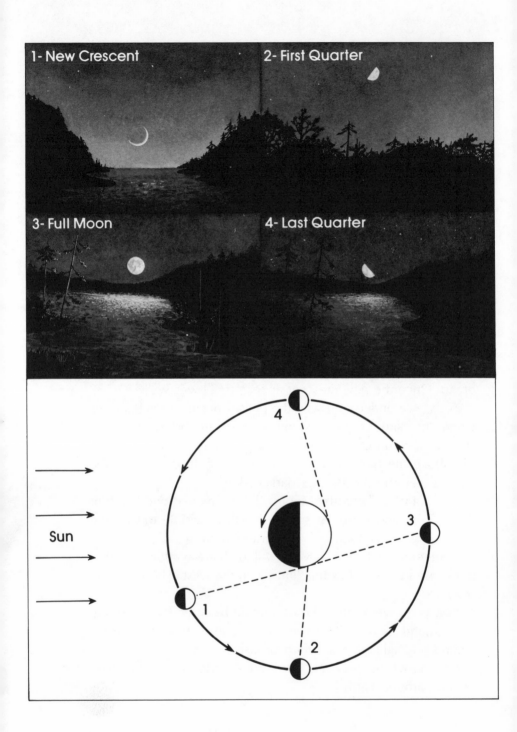

1- New Crescent

2- First Quarter

3- Full Moon

4- Last Quarter

Sun

If you watch the Moon from night to night throughout the year, you'll notice another lunar motion. The next time the Moon is full, watch for moonrise, which takes place at sunset. Make a drawing of the horizon and put a mark and the date at the location where the Moon appears. Keep the record for a year, noting especially the December-January and June-July locations. You'll be amazed by the range of the rising positions, from far south of east to far north of east.

◄ *The new crescent moon (1) appears in the western sky shortly after sunset.*

A week later, the first quarter (2) is seen in the southern sky at sunset.

After two weeks, the full Moon (3) is seen in the east as the Sun sets in the west. The Moon is opposite the Sun.

The last quarter Moon (4) rises about midnight. As the days go by, it diminishes to a crescent, rising closer and closer to the time of sunrise.

The full Moon is opposite the Sun. In winter, when the Sun is low in the sky, the full Moon rides high.

In summer, when the Sun is high, the full Moon rides low. It follows a short path across the sky.

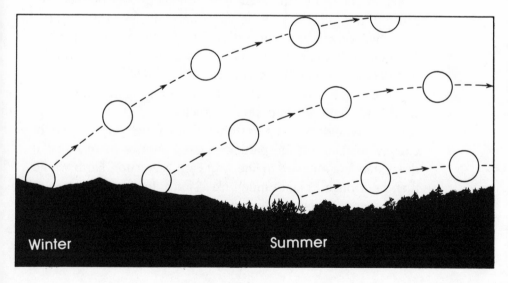

Winter   Summer

The full Moon is opposite the Sun, not only east for west, but also south for north. For example, in summer, the Sun is north of east. The full Moon is south of east, and it follows a short path across the sky. During winter, the Sun is south of east. The full Moon is far to the north. It follows a long path from east to west. The difference is quite unbelievable.

What will eventually happen to the Moon?

No one really knows. However, far in the future, the Moon may be pulled apart and broken into small pieces by Earth's gravity, if it isn't first swallowed by the Sun.

Long before the Moon breaks up, it will move farther from Earth. The sequence of events will require billions of years— moving away from Earth and then moving in closer. Right now, the Moon moves away about 2.5 centimeters a year.

Because of the drag of water as it is pulled by the Moon, tidal action slows down Earth's rotation about one second in fifty thousand years. As Earth slows, the Moon spirals away from us. After billions of years, Earth will slow so much, it will take 55 days to make one rotation.

Billions of years later, opposite conditions will prevail. Because of the attraction of the Sun, Earth will speed up and the Moon will slowly spiral inward. Eventually, the Moon will come in so close that Earth's gravity will be extremely strong on the Earthside of the Moon and considerably weaker on the far side. These unequal forces would pull the Moon apart. The Moon would be shattered into small particles of dust and ash. The particles would become distributed in a belt around Earth. Gravity working on the particles would arrange them in a flat disk extending outward in line with Earth's equator. Earth would then be a ringed planet much like Saturn, Jupiter, and Uranus.

This series of events would require billions of years, much longer than the five billion years the Sun is expected to shine

Eventually, the Moon may move close enough for Earth's gravity to pull the Moon apart and shatter it into small particles. These particles would then form into a ring encircling Earth, just as rings surround Jupiter, Saturn, and Uranus.

with its present brilliance. A sudden increase in solar temperature may cause the inner planets and their satellites to vaporize. They may become surrounded by hot solar gases. Much of the solar system may have disappeared long before the Moon could have come close enough for it to be reduced to a ring of ashes.

Among the satellites, the Moon's size is impressive—one-fourth the diameter of its planet. This is not true of the satellites of Mars, the next planet from the Sun.

3 PHOBOS AND DEIMOS —THE MARTIAN SATELLITES

How were the satellites discovered?

Outward from Earth is Mars, with its two small satellites, Phobos and Deimos. The words *phobos* and *deimos* are Greek for *fear* and *panic*. They are fitting names, for the satellites are the companions of Mars, and Mars is the Roman god of war. The planet was given the name because of its bloodred color.

The satellites were named by Asaph Hall (1829–1907), an American astronomer who discovered them in 1877 while at the Naval Observatory in Washington, D.C. They had escaped earlier discovery because they are very small and are located very close to Mars. Through even a large telescope, they appear as tiny dots—when conditions are good. Most often, they are not visible because they are in the glare of sunlight reflecting off Mars itself.

Johannes Kepler, the German astronomer who lived in the early 1600s, had suggested that Mars had two satellites. He reasoned that since Earth had one satellite and Jupiter had four, the planet between Earth and Jupiter should have two. Now we know that such reasoning was not sound, for there is no order in the number of satellites of the planets.

Asaph Hall was able to make the discovery in 1877 for two reasons—he had a large telescope to work with, and at that time, Mars was especially close to Earth. During that same period, the Italian astronomer Giovanni Schiaparelli reported that there were *canali* on Mars. This is the Italian word for *channels,* formations that could have been made naturally. However, the world preferred to think that Schiaparelli had seen canals on Mars, structures that could be made only by some kind of intelligent creatures. His report stimulated thinking about Martians and the possibility that the planet supported civilized people. Today few people have such a belief. But many believe that one day we'll discover some lower forms of life on the planet.

How would the satellites appear from Mars?

Phobos and Deimos are both close to Mars. The orbit of Phobos is 9300 kilometers from the Martian surface, while Deimos moves at a distance of 23 490 kilometers.

At these distances, Phobos would appear one-third the size of our full Moon. Deimos would be much less noticeable; it would appear as little more than a fairly bright star.

Phobos moves very rapidly, completing a journey around Mars in less than 8 hours. That's much faster than it takes Mars to rotate. As a result, Phobos rises in the west and sets in the east.

The period in which Deimos revolves is 30 hours, which is longer than the 24 hours 37 minutes it takes Mars to rotate. Therefore, Deimos would rise in the east and set in the west, just as does Earth's moon.

Phobos (top), *the larger of Mars' satellites, has huge craters.* ▶
Viking Orbiter 1 flew within 480 kilometers to obtain the photos in this mosaic. A close-up of the satellite (bottom) *reveals a surface much like that of the highland regions of the Moon.*
NASA

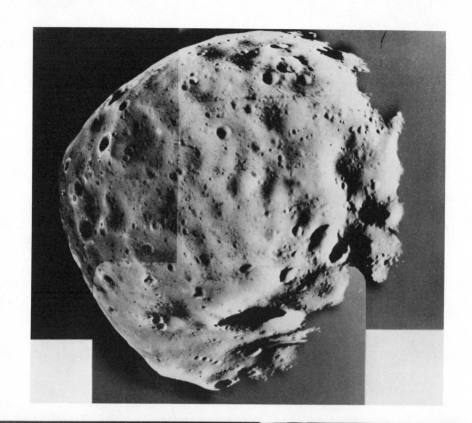
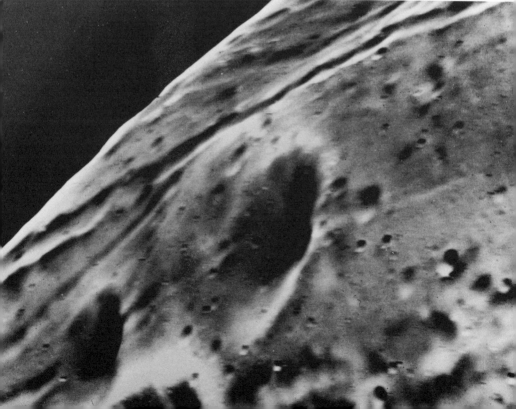

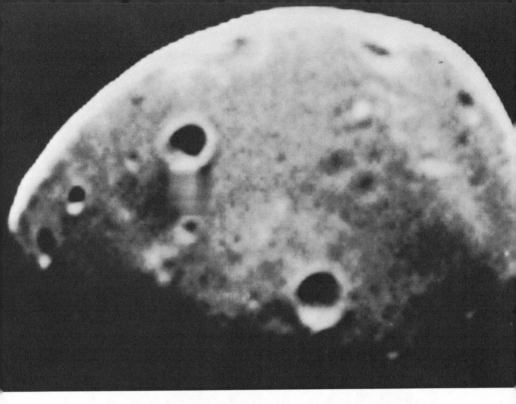

Deimos (top) *is also cratered. Separate boulders the size of a house can be seen in the close-up view of the satellite (bottom), taken from only 50 kilometers away.* NASA

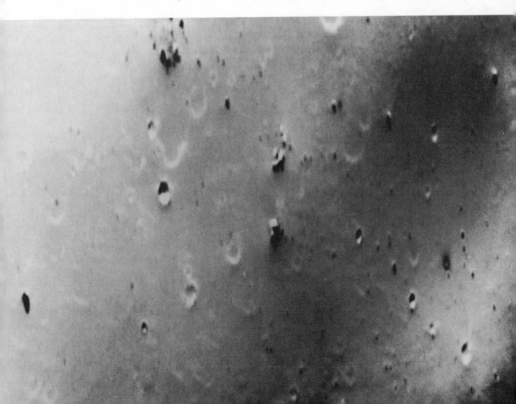

Could the satellites be asteroids?

Since the discovery of the satellites, people have known they were small. However, actual measurements were not possible until 1971, when photographs were made by Mariner 9, a planet probe that moved in close to Mars. Both satellites are shaped like baking potatoes; they are ellipsoids. Their sizes are given in three dimensions rather than simply one diameter, as would be the case if they were spherical.

Phobos, the larger of the two, measures 27 by 22 by 18 kilometers. Deimos is about half as large. It measures 16 by 12 by 10 kilometers. Both satellites are completely barren, desolate worlds that are covered by craters. Some two billion years ago, they must have been bombarded heavily by rocks and boulders flying through space. Perhaps the satellites were once a single larger object that broke apart.

The surfaces of both satellites are unusually dark, making them appear much like asteroids. It is entirely likely that the satellites are asteroids that were captured long ago by the gravity of Mars. Their orbits may have intersected that of Mars, and so the capture became possible.

The satellites have little, if any, effect on Mars. Deimos would attract little notice as seen from the Martian surface. However, Phobos would be rather interesting—large enough to be noticeable, reappearing in the west every 7 hours and 40 minutes.

The satellites of Jupiter, the next planet beyond Mars, are much more impressive, especially the four larger ones.

 # JUPITER'S
SATELLITES

How many satellites does Jupiter have?

Jupiter is the largest of all the planets; it also has the strongest gravity. It is able to hold a large number of satellites. Sixteen have been detected and recorded. There may also be additional satellites that not have not yet been discovered.

Some of Jupiter's satellites are only a few kilometers across. All are under 270 kilometers across, except the four larger ones —Io, Europa, Ganymede, and Callisto. Their names are taken from mythology. Jupiter was the greatest of the gods and so had a good many followers. These four satellites, as well as the other Jovian satellites, were named after followers of the god Jupiter.

How did the Galilean satellites establish a belief?

Until 1610, the only satellite that had been seen in the solar system was the Moon. In that year, the Italian astronomer Galileo Galilei, using one of the telescopes he had built, discovered four major objects moving around Jupiter. Ever since then, they have been called the Galilean satellites.

This was an especially exciting discovery, for it came at a time when people were not yet willing to accept the Sun-centered solar system. They still believed that Earth was central and all other objects moved around it. In 1543, Nicolaus Copernicus, the Polish astronomer, said that the planets all moved around the Sun. Earth was one of those planets, he said, and was not the center of the solar system. However, he could not present any observation to prove his ideas. But Galileo could, and he did.

When he first observed Jupiter, Galileo saw three of the satellites. In following nights, a fourth appeared. Night after night, he watched the four satellites and saw them change position; they moved from one side of the planet to the other. The satellites were going around Jupiter. So all objects did not go around us; Earth was not the central object of the solar system. The planets were satellites of the Sun; they moved around the Sun, just as the Galilean satellites moved around Jupiter.

Using a small telescope, or binoculars if you can hold them steady, you can see the satellites and chart their positions, just as Galileo did almost four hundred years ago.

From an almanac, you can find when and where Jupiter is visible. Once you have located it, look carefully for small dots on either side of the planet. Chart their locations. On following nights, make additional charts and you will be able to see how rapidly the satellites change positions. The times required for the satellites to move around Jupiter are shown below:

| | ORBITAL PERIOD | | DIAMETER |
	DAYS	HOURS	KILOMETERS
Io	1	18	3630
Europa	3	13	3138
Ganymede	7	3	5262
Callisto	16	16	4800

Should you happen to see all the satellites at once, it will be easy to identify each one of them by its distance from the planet. The nearest is Io, which is some 349 000 kilometers away, and the farthest is Callisto, which is at a distance a bit over 1 600 000 kilometers.

What are the bulges on Io?

In 1979, two Voyager planet probes flew by Jupiter. They took thousands of photographs of the planet and of many of its satellites and sent them back to Earth. Scores of people studied those photos. One of them was Linda Morabito, at that time a flight engineer at the Jet Propulsion Laboratory in California. She noticed a small knob or bulge on one of the photos of Io and wondered what it could be.

Gases thrown out by an erupting volcano appear along the edge of Io (left). After the Voyager 1 discovery of the first volcano on Io, a total of twelve active volcanoes were found. The photo on the right shows two volcanic eruption plumes about 100 kilometers high. JET PROPULSION LABORATORY

Closer inspection revealed that it was an umbrella-shaped formation of gases spouting from the surface—a volcano. When other pictures were studied, twelve active volcanoes were seen. Io is the only place in the solar system, other than Earth, that is known to have active volcanoes—although they may exist on Enceladus, one of Saturn's satellites. Apparently, Io is in the volcanic period of its evolution.

Why does Io spout sulfur?

Unlike Earth's volcanoes, which eject molten rock and ash, those of Io spout sulfur and sodium. The material is thrown at least 240 kilometers above the surface at a speed somewhat greater than 900 meters a second. (Earth's volcanoes eject ash and gases at speeds around 135 meters a second.) Io has low gravity and practically no atmosphere, so the debris forms a high, evenly shaped umbrella and rains down onto the surface in a smooth dome shape.

The sulfur must come from beneath the surface, where it is very hot, liquid, and gaseous. Rocks and sulfur are heated and vaporized, and pressure pushes the molten material up through vents in the surface. The lava spreads over the surface for hundreds of kilometers. Sulfur changes to dark crystals, and so the volcanoes of Io are surrounded by dark sulfur lakes. The temperature in the lakes is about 90° Celsius (C).

Craters on Io may have been formed by the impact of meteoroids, but the craters have been worn down or filled in by lava. The surface now has no craters on it; Io is among the smoothest of all the satellites.

Why is Io so hot?

Io is not only smooth, it is also hot. Its heat cannot come from the Sun, for the satellite is too far away to receive very much. Rather, the heat probably comes from tidal effects. Jupiter's other three large satellites produce strong tides on Io, raising and lowering the surface as much as 110 meters. The alternate up-and-down movement produces heat in much the same manner as a wire gets hot when you bend it back and forth.

Beneath the surface, there may be a huge reservoir of molten sulfur. The surface itself is orange red with splotches of yellow and black. Here and there, one sees white patches of sulfur snow. Io looks like a king-size, spherical pizza.

Io has a thin atmosphere. Other satellites with atmospheres are Titan, which is Saturn's largest, and Triton, one of the satellites of Neptune. Io's atmosphere glows with light emitted by sodium atoms. The sodium streams over 160 000 kilometers along Io's orbit, forming a partial ring around Jupiter.

What is Jupiter's torus?

Io moves in a strong electric field, which is another mystery. Like Earth, Jupiter is magnetic. Its magnetic field extends far into space. Io moves through it, setting up an electric charge of five million amperes. By comparison, the current in a hundred watt bulb is only one ampere.

Many of the gases and dust particles thrown out by Io's volcanoes escape and are captured by Jupiter. They are held in a tube shape, or torus, that surrounds the planet. Io moves inside that torus of charged particles. Violent lightning storms occur continually on Jupiter, as tremendous electric charges are released. Storms probably also occur on Io.

What is it like on Io?

Io is probably as old as the rest of the solar system. As mentioned earlier, it was cratered at one time, but the craters can no longer be seen—they are filled in. Enough sulfur is thrown out by the volcanoes to cover the surface completely in one million years, which is only a moment compared to the age of the satellite.

The overall temperature of Io is about 150°C below zero. But there are occasional hot spots. Lakes of sulfur are hot, for example, as are the dark patches on the satellite. These are probably sulfur lakes covered by a thin crust of solid sulfur. Underneath that crust, the temperature of the dark splotches is 600°C, hot enough for sulfur to be liquid.

Beneath the surface of Io, sulfur seas may extend around the satellite. Perhaps the surface of Io is completely covered by new sulfur that has welled up from the interior. Sulfur rain and snow fall from the tremendous explosive fountains covering the landscape, adding to Io's incredible desolation.

Io is a dreaded place. It fascinates astronomers and geologists, who are challenged to find answers to its many mysteries.

Why does Jupiter have a ring?

When the Voyager probes photographed Jupiter, they discovered that a ring surrounds the planet. The ring seems to be a system of three separate bands. The formation was a great surprise. Why should it be there? The outer, brightest part is about 800 kilometers wide, while an inner formation is much dimmer and wider —perhaps 6000 kilometers. A very broad band lies inside this. The three rings appear to be only about 30 kilometers thick.

The ring, or rings, contain very small particles, each separate from the others and each in its own orbit. So one could say that Jupiter has billions of satellites, as many as there are particles in the ring. While most of the particles are very small, there may be large chunks as well. In addition, two small "ring" satellites, Adrastea and Metis, are embedded in the formation.

Exactly why the ring exists is still a mystery. However, it is probable that Io's explosive volcanoes have a lot to do with the ring. It is very likely that the volcanoes eject sulfur and sodium which are then captured by Jupiter.

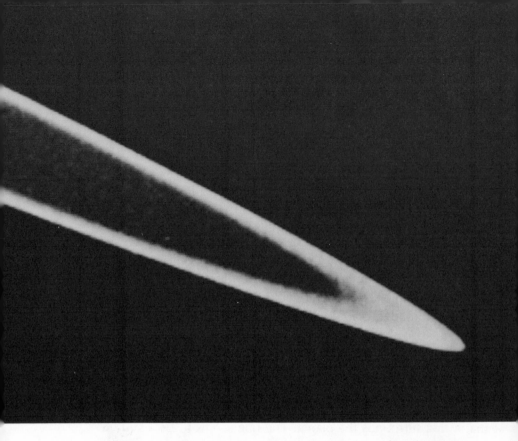

Rings of small particles surround Jupiter. Seen within the inner edge of the brighter ring is a fainter ring which may extend all the way down to Jupiter's cloud rings. JET PROPULSION LABORA-TORY

The number of ringed planets is now three—Saturn, Uranus, and Jupiter. Some astronomers suspect that Neptune also has rings, although there is no reliable evidence that they exist.

Why is Europa so smooth?

The second of the Galilean satellites looks like a round, cracked egg. Its surface is smooth, broken by long, irregular dark lines. The bright areas are water ice, and the lines may be cracks that have been filled in with debris.

Like most other satellites, Europa must have been bombarded

billions of years ago during early stages of its formation. However, there are few if any craters there now, meaning that the rims of craters have sunk or have been worn down. The cracks are only a few feet deep, making the surface of Europa unusually smooth. The cracked surface probably means that it is thin and overlays a hot interior. The heat is probably generated by tidal forces, much as heat is produced on Io.

The very smooth glasslike surface the satellite once may have had has been cracked by these tidal forces. In much the same way, here on Earth, ice in tidal waters is broken into chunks, and great masses pile one on the other, with long cracks between them. It is possible that the cracks of Europa appear black because warmer material from inside the satellite was pushed up through them, cooling as it reached the surface.

The complex patterns on Europa's surface suggest that its icy covering has been fractured. JET PROPULSION LABORATORY

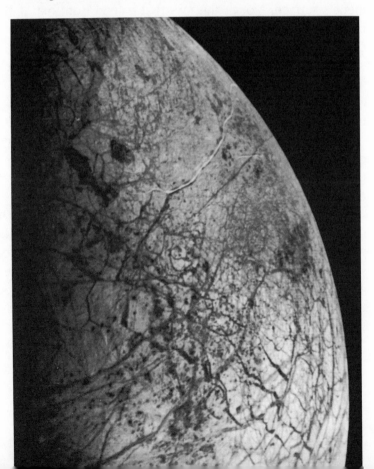

Has Ganymede changed?

Ganymede, with a diameter of 5262 kilometers, is the largest of Jupiter's satellites. Indeed, it is the largest satellite in the solar system; it is also considerably larger than the planets Pluto, with a diameter of 3000 kilometers, and Mercury, 4878 kilometers.

Like the other Galilean satellites, the time required for Ganymede to go once around Jupiter equals the time it takes for the satellite to rotate. As a result, the same face of the satellite is always turned toward Jupiter.

Voyager probes revealed a surface that is cratered and grooved. The craters on much of the surface have not been worn down or filled in and so are much the same as when they were first formed—perhaps four billion years ago.

None of the craters are very deep. There are also no high mountains on Ganymede. Perhaps its surface has been sinking, softening somewhat so the mountains sink into it. This seems reasonable, for water ice is believed to be a large part of Ganymede's surface. Notice the rays that extend from many of the craters. These are believed to have been caused when boulders crashed into ice formations.

Photos of Ganymede show multitudes of shallow grooves, like tracks in sand. There is a pattern of light ridges and darker grooves, extending 500 or 600 kilometers. One also sees places where sections of the surface appear to have pulled apart.

The density of Ganymede is 1.9, not much greater than the density of water. This means the satellite is about half water and half rock. (The high density of Earth means a much greater percentage of our planet is solid rock.)

The surface of Ganymede is by no means uniform. There may be craters atop grooves or grooves atop craters. Mixtures of various formations indicate the satellite has gone through several periods of impact. Surface structure implies that the last great impact occurred at least three billion years ago. It seems the surface has not changed a great deal since then.

Craters, ridges, and tracks on the surface of Ganymede indicate many stages in the formation of its surface. The tracks may be ridges and gulleys. The presence of a variety of features implies that the satellite is very old.   JET PROPULSION LABORATORY

How did Callisto's mountain rings form?

Callisto is the outermost of the Galilean satellites, at an average distance of 1 770 000 kilometers from Jupiter. It circles the planet once in 16 days, 16 hours. Like Ganymede, it has a low density, and so is probably half water and half rock.

Of all the satellites, Callisto has the record number of craters. Some are large, others quite small, all are shallow—none more than 100 kilometers deep. At one time, the craters were probably deeper. However, since so much of the surface is ice, it is able to flow slowly and fill in the lower depths. A few of the craters appear to be filled with ice.

A most unusual feature of Callisto is a large impact area about 500 kilometers across. This area is surrounded by two dozen mountain rings that extend some 3000 kilometers from the center. The mountain rings may be waves that were quickly frozen. Long ago, there may have been a tremendous collision with a free-flying asteroid. The heat that was generated may have melted ice over a wide area and caused huge waves to spread outward from the point of impact.

The temperature on Callisto is very low, at least 200°C below zero. At that temperature, the waves would have frozen instantly, preserving their shape. The rings of mountains may be largely water ice with rocky material imbedded in it.

When the collision occurred, craters in the region were filled in or eroded away, leaving a rather crater-free impact zone.

Could any of Jupiter's satellites be asteroids?

Io is the innermost of the Galilean satellites. But Amalthea, a much smaller satellite, travels in the space between Io and Jupiter. It takes Amalthea only 12 hours to complete an orbit, meaning it moves about 96 500 kilometers an hour.

Of the lesser satellites, only Amalthea has been photo-

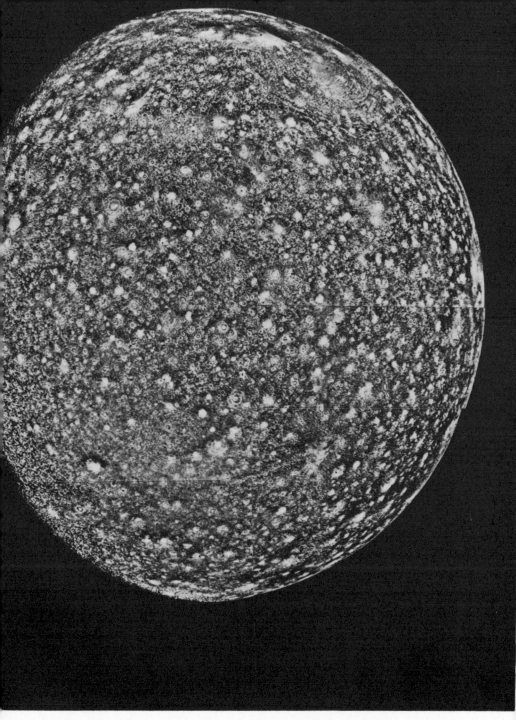

This photomosaic of Callisto, composed of nine frames, shows the very uniform crater distribution across the satellite. The bright rayed craters are probably quite young. *JET PROPULSION LABORATORY*

graphed closely enough to show details. There appear to be craters on its surface. It is elongated like a potato and is about ten times larger than Phobos. While the satellite goes around Jupiter, it makes a single rotation. Therefore, like the Galilean satellites, Amalthea keeps the same face toward the planet.

Because of the shape and size of Amalthea, it is reasonable to suppose that the satellite was originally an asteroid. Jupiter is at the outer edge of the asteroid belt, the region that lies between the orbits of Mars and Jupiter. Jupiter also has strong gravity, and so it could capture asteroids and pull them out of orbit.

Two other small satellites—Metis and Adrastea, which we mentioned earlier—and possibly a third, lie between Amalthea and Jupiter. However, they have not been photographed close up, and so we have little information about them.

Perhaps all the lesser satellites are captured asteroids. Indeed, Jupiter may still be pulling asteroids out of orbit and so adding to its satellites. Presently, the total stands at sixteen, although all of these are not confirmed. On the other hand, there may be others that have escaped observation.

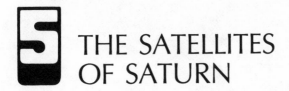

THE SATELLITES
OF SATURN

How many satellites are there?

You cannot see Saturn's rings with your eyes alone. But even a small telescope shows that the planet is spectacular. In the early 1600s, Galileo discovered that it was different from the other planets. Noticing that there were bulges on either side of it, Galileo said that Saturn had ears.

He was never able to see the bulges clearly, and so he was unable to explain them. But a few decades later, in 1657, the Dutch astronomer Christian Huygens was able to see the bulges more distinctly. It was apparent to him that Saturn has rings. He was unable to determine what they were made of, though, or to discern details of their structure.

Recently, space probes have moved in close to the planet and revealed that Saturn has hundreds of rings. They are made of separate particles; most are very small, but some are as large as a small car. Some of the rings are wide, others narrow, and all are thin.

We mention the rings because, in a way, each of the particles in the rings is a separate satellite; it is in orbit around the planet. From that standpoint, one could say that Saturn, like

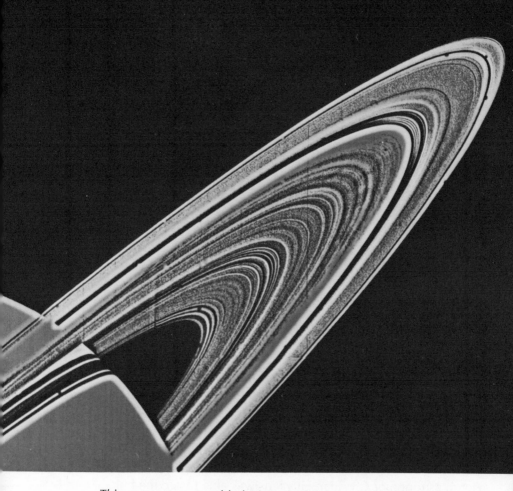

This computer-assembled picture of Saturn's rings shows approximately ninety-five individual concentric features. The ring structure, once thought to be produced by the gravitational interaction between Saturn's satellites and the orbit of the ring particles, has now been found to be too complex for this explanation alone. NASA

Jupiter, has billions of satellites. However, it would be impossible to identify each particle, determine its orbit, and understand its characteristics.

When such yardsticks are applied, it turns out that the number of Saturn's satellites is much smaller. In our table on pages

6 and 7, we show seventeen. That's the official number. Some observers count twenty-two. Eventually, more satellites may be charted. Long exposures made with cameras aboard the Hubble Space Telescope may reveal them.

Could there be life on Titan?

Saturn was a Roman god, the same as Chronos, an earlier Greek god. The word *chronos* means time. Its name is appropriate because, of all the planets known in the days of the early Greeks and Romans, Saturn takes the longest time to move across the sky. Titan and the other Saturnian satellites—including Tethys, Calypso, and Dione—were named after companions of the god.

This montage combines individual photos taken by Voyagers 1 and 2 during their Saturn encounters. Clockwise, from upper right, the satellites are Titan, Iapetus, Tethys, Mimas, Enceladus, Dione, and Rhea. NASA

Titan, with a diameter of 5150 kilometers, is the largest of all Saturn's satellites; it is just a bit smaller than Ganymede and considerably larger than the Moon.

Titan has an atmosphere which is the deepest of all the satellites'. It is so dense that pressure at the surface is one and half times that on Earth at sea level. The atmospheric gases appear to be mostly nitrogen, methane, ethane, argon, and traces of cyanide. Methane and ethane are hydrocarbons, meaning that the molecules are made of combinations of hydrogen and carbon. They both burn and are used as fuels here on Earth. They are both poisonous to living things, as is cyanide.

Because of this, Titan could not support life, and it would not be a very pleasant place to visit. The satellite is also intensely cold—some 250°C below zero.

It is very likely that the surface of Titan is covered with liquid gases. (When gases become very cold, they change to liquids. When even colder, the gases may freeze solid.) The liquid ocean of methane, ethane, and nitrogen may average half a mile deep. Beneath that layer, there may be rich deposits of acetylene, another hydrocarbon.

Saturn is surrounded by an energetic radiation belt; a region that has a magnetic field strong enough to attract radiation particles ejected from the Sun. Titan is at the outer edge of the belt and so is relatively free of radiation particles. Titan goes around Saturn once in 16 days. Saturn's gravity causes Titan to rotate in the same length of time. This means the same face of Titan is always turned toward Saturn. The satellite is probably misshapen, bulging on the Saturn side and somewhat depressed in the polar regions. Density indicates that the satellite is, for the most part, ice and gases.

What are the lesser satellites made of?

All the other satellites of Saturn are small—some only a few kilometers across, none more than 1530 kilometers in diameter.

The satellites appear to have low densities, implying that they are made mainly of ice.

Photographs of Enceladus cause some observers to suspect that active volcanoes are erupting. They are not as obvious as those on Io, but they may exist. Furthermore, sections of the surface of the satellite seem to be free of craters, indicating there has been filling in of holes, perhaps with lava. The surface of Enceladus may be new and quite different from the way it was when the satellite formed.

Others argue against that possibility. They point out that the interior of Enceladus is not very hot, for it is not subject to tidal action to the extent that Io is. Remember, one explanation for the volcanoes of Io is the tidal heating of the satellite.

The other satellites are presently at locations that do not cause severe tidal changes on Enceladus. It is suspected, however, that some ten million years ago the satellite Janus may have been closer. Indeed, it may have been a larger satellite that was split into smaller pieces by the strong gravity of Enceladus.

Hyperion is a small satellite that is fairly close to Titan. Therefore it is affected strongly by the pull of the giant satellite. Some people believe the effect is so great that Hyperion's rotation rate is very erratic. It may spin once in 21 days, the same as its orbital period. Or it may spin twice as fast. At other times, it may not spin at all. Many more photos of the satellite are needed in order to find the answer to this mystery.

Iapetus, a bit under 1500 kilometers across, is located some 6 000 000 kilometers from Saturn. Curiously, one half of the satellite reflects a lot of sunlight, while the other half is a poor reflector. In the picture on page 52, the entire disk is lighted by the Sun, showing clearly the difference in reflectivity. It is as much as the difference between a surface covered with snow and one paved with asphalt.

The reason for this is not known. However, it has been suggested that widespread volcanoes may have erupted over part

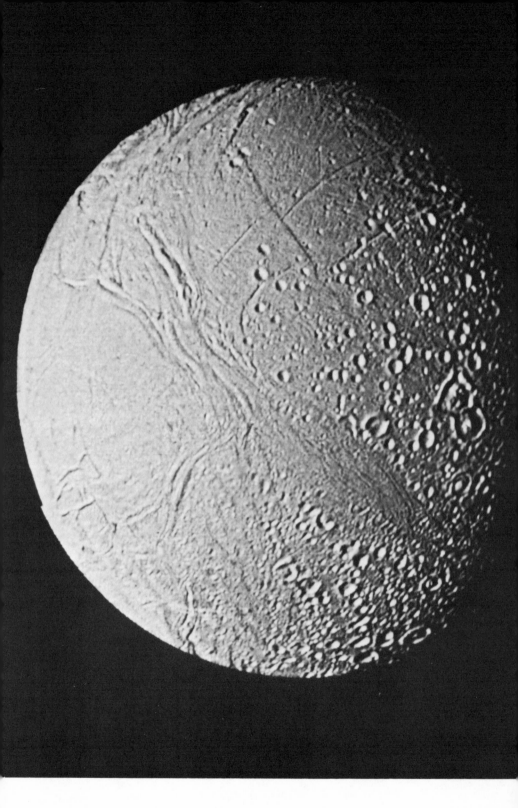

of the satellite, spreading dark lava over half of it. The idea is not widely accepted, though, because it does not explain why there should be such a sharp division between the two regions.

Many people feel it is more reasonable to connect the satellite to Phoebe, the outermost of Saturn's satellites. It is possible that Phoebe was bombarded by objects massive enough to break up the surface and eject debris into space.

The particles would have been attracted to Iapetus by the satellite's gravity. And they would have accumulated on the leading half. This is because Iapetus rotates and revolves in the same period: one rotation in the time required for one revolution. The "front" half would tend to sweep up the debris. As the material collided with Iapetus, lighter ice compounds would have been thrown off. Gradually, the surface would have become covered with darker rock material.

Perhaps this is so. The light reflected from Phoebe, however, does not match that reflected from the dark side of Iapetus. There could be an explanation, say some astronomers. In moving from Phoebe to Iapetus, the particles might have been changed somewhat and so may present a slightly different appearance.

Moving inward toward Saturn, the next satellite is Hyperion. One would expect that half of it would also be dark and the other half bright. That would be true if the satellite rotated once while completing a revolution. If it did not, the darker material from Phoebe would have spread evenly around Hyperion, making the entire satellite rather dark. And that is how it appears. Also, when astronomers studied the dark areas of both satellites, they found them to be identical in structure. It seems that both originated in the same manner, very possibly because of accumulation of debris from Phoebe.

◄ Parts of Enceladus are crater-covered, while other regions contain extensive ridges and grooves and are free from craters. Enceladus resembles Jupiter's moon Ganymede, which is, however, about ten times larger. JET PROPULSION LABORATORY

Craters appear in both the dark and bright regions of Iapetus.
JET PROPULSION LABORATORY

How were the rings of Saturn produced?

As mentioned earlier, the rings are made of separate particles. In some of the rings, the particles are close together, making the rings dense and so able to produce sharp shadows on the planet, as shown in the photo on page 46.

No one knows the complete story of the rings. It's possible that at one time there were no rings—only separate satellites. As a satellite moved closer to the planet, gravity on the planet side of the satellite would have been much greater than on the far side. Since the force was uneven, there would have been great stress on the satellite, enough to pull it apart and shatter it into a cloud of particles. Because of their motion around the planet and because of gravity, the particles would have become arranged in a disk—the rings of Saturn.

Or possibly the particles never combined to form a satellite. High velocity of the particles prevented them from joining together, and so they have remained spread around the planet.

For the most part, the particles appear to be water ice. There are also rocky particles that are covered with ice. Although there are hundreds, even thousands of rings, they add up to very little mass—perhaps only one-millionth the mass of the Moon.

 # THE SATELLITES OF URANUS

In mythology, Uranus is the oldest of the gods, the ruler of the world. The major Uranian satellites—Miranda, Ariel, Umbriel, Titania, and Oberon—are named after the mythological followers of Uranus.

All of the satellites are rather small, ranging from Miranda, about 500 kilometers across, to Oberon, which is 1630 kilometers in diameter. Miranda, the closest to Uranus, was discovered in 1948. Ariel and Umbriel have been known since 1851. The other two, Titania and Oberon, were identified by Sir William Herschel, who, in 1781, discovered Uranus itself. Ten more satellites were discovered by Voyager 2 in January 1986.

The time the satellites take to go around the planet and their distances from Uranus are:

| | ORBITAL PERIOD | | DISTANCE |
	DAYS	HOURS	KILOMETERS
Miranda	1	10	135 000
Ariel	2	12.5	193 000
Umbriel	4	3.5	257 000
Titania	8	17	442 000
Oberon	13	11	595 000

Why do the satellites seem to reverse direction?

Powerful telescopes are needed to see the satellites. Over a period of years, astronomers have noticed that the satellites appear to move clockwise at one time and, years later, to move in the opposite direction. This is because Uranus spins on its side. The axis of Uranus lies almost on the same line as its orbit around the Sun. The planet takes eighty-four years to complete one journey. During half that time, the North Pole of Uranus is toward the Sun. For the next forty-two years, the South Pole is in that position. The satellites are located about halfway between the poles. Therefore, when either pole of Uranus is turned toward the Sun, we see the satellites moving in circles around the planet, first in one direction and then in the other.

When sunlight is falling on the equator of Uranus, the satellites appear to move up and down, rather than in circles—as shown in the drawing on this page.

Because the axis of Uranus is tilted so much, the poles are pointed toward Earth at intervals of some forty years. When one pole is toward us (as in 1946), *the satellites appear to go clockwise; at the opposite position* (1985), *the satellites appear to move counterclockwise.*

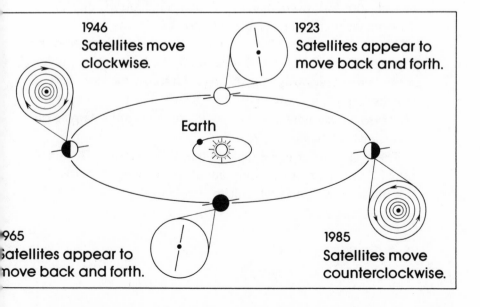

1946
Satellites move clockwise.

1923
Satellites appear to move back and forth.

Earth

965
atellites appear to nove back and forth.

1985
Satellites move counterclockwise.

All the satellites have low densities, only slightly more than that of water. This leads astronomers to suspect they are made largely of water ice. Umbriel has many large, overlapping craters, but the satellite appears to be old and inactive. The other four major satellites have much evidence of internal activity, such as deep fractures, glacial flows, and long ridges.

How were the rings of Uranus discovered?

Like Saturn, Uranus has a system of rings. These rings are composed of blocky chunks of dark material that orbit the planet about once every eight hours. The gaps between the chunks are filled with discrete particles. One could say that Uranus, like Jupiter and Saturn, is surrounded by billions of satellites.

The rings were discovered by accident in 1977. In that year, astronomers in a high-flying airplane equipped with a telescope were observing Uranus. The planet was to move in front of a star. Observers hoped there would be a dimming of the starlight which would give them information about the atmosphere of Uranus. However, the astronomers were puzzled because the starlight dimmed long before the atmosphere layer of Uranus was in front of the star. It dimmed, then brightened, then dimmed five times before the star was behind Uranus. Later investigations showed that the dimming occurred nine times. This meant there were nine rings around Uranus.

Then, in January 1986, a tenth ring was discovered by Voyager 2. This new ring lies just inside the brightest and outermost of the previously known rings. Although the discovery of the first nine rings was made by chance, the most recent discovery is an example of how new instruments are enlarging our knowledge of space.

The rings appear to be divided into three groups. The smallest of them seems to be only about 4.5 kilometers across. All the rings are dim and difficult to detect.

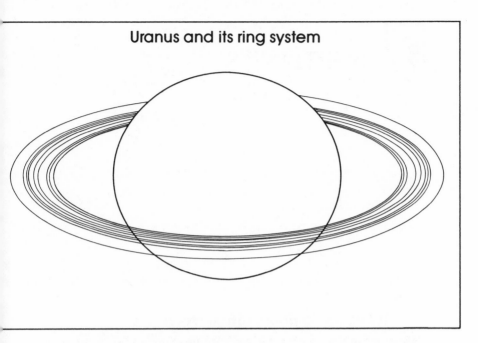

Uranus and its ring system

Quite by accident, nine rings were found to be associated with Uranus. The material in the rings, mostly ice and rock, may be debris from a moon or moons that disintegrated when Uranus was young.

What are some of the discoveries made by Voyager 2?

Photographs from the spacecraft have revealed a bright, dough-nut-shaped ring orbiting Umbriel, the darkest satellite. Also shown in photographs are valleys on the surface of Ariel. Some are filled with bright deposits of material, others seem to have glaciers spread across dark plains. A deep fault line runs across Titania, reminding scientists of the Great Rift Valley of East Africa. Many more exciting discoveries are expected to be made by the spacecraft, which was launched in 1977.

7 NEPTUNE'S SATELLITES

Why was Neptune discovered?

Neptune moves in an orbit far from the Sun, so far that sunlight takes more than five hours to reach it. (Sunlight takes about eight minutes to reach Earth.) Neptune is fairly large. However, because of its great distance from the Sun, it appears very dim, and so it was not discovered until 1846.

Astronomers computed the orbit of Uranus and were able to predict its future locations. But the planet did not behave properly; Uranus was deflected somewhat from its predicted course. This meant there had to be a massive body out beyond Uranus, one that had enough gravity to affect the motions of Uranus. Astronomers predicted where such a body should be located. Neptune was discovered in 1846 when telescopes were pointed to that position.

How were Triton and Nereid produced?

In that same year, Triton, the larger of Neptune's two satellites, was seen. Nereid, the outermost satellite, is much smaller. It was

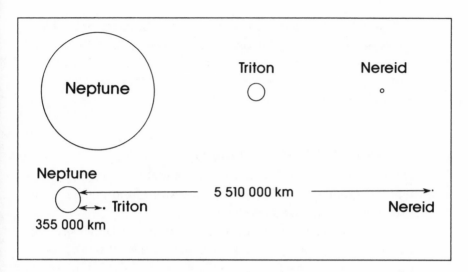

Neptune has two satellites. Triton, the larger by far, is the one nearer to the planet. Tiny Nereid is over 5 500 000 kilometers away.

not detected until 1949, more than one hundred years later. In mythology, Triton and Nereid are minor gods attending Neptune, god of the sea.

Nereid is a small satellite, only about 400 kilometers across, and it is located some 5 500 000 kilometers from Neptune. Possibly, it was an asteroid that once moved in close to Neptune and was captured by the planet. Or it may be debris from a major collision with a massive object.

Triton, with a diameter of 3500 kilometers, is among the very large satellites—one of the seven that have diameters greater than 3100 kilometers. It revolves around Neptune once in about 5 days, moving from east to west. That is backward—or, more properly, retrograde—since most satellites revolve from west to east around their planets. As it goes around Neptune, the same face of Triton is always turned toward the planet. Like the Moon, the rotation period of Triton equals its period of revolution.

The satellite appears to have a rocky surface, with liquid methane filling depressions in it. Above the surface, there is an atmosphere of methane. Perhaps other gases, such as ethane and cyanide, also exist there, although as yet they have not been identified.

Does Neptune have rings?

A few astronomers think so. You recall that the rings of Uranus were a serendipitous discovery—they were found while scientists were looking for something else.

Edward Guinan, an astronomer at Villanova University, believes that he made a similar discovery while observing Neptune.

In 1968, he and other researchers were in New Zealand watching Neptune as it moved in front of a star. Variations in brightness were recorded on punch cards, the kind that are fed into a computer for analysis. Unfortunately, the cards got wet during the boat trip from New Zealand. This caused them to warp, and so made it impossible to put them through a card reader.

Guinan set them aside and was not reminded of them until 1977, when the rings of Uranus were discovered. The cards were still warped so he and his students set about copying the cards, one at a time. Five years later, in 1982, they were finished. When the cards were fed into a computer, a partial cutoff of light showed up about 5000 kilometers above the surface of the planet. A similar dimming was seen as the planet moved away. The implication is that a ring caused the dimming.

The finding is not widely accepted, however, for other astronomers have not been able to detect the dimming. So whether or not Neptune has a ring is still a mystery. Some people say it is questionable that a ring could develop, or be maintained once it formed, because of Triton's mass. According to these observers, the gravity of Triton would be great enough to disrupt the motions of ring particles, causing the breakup of such a formation.

Considerably more evidence is needed before astronomers will say there is a ring around Neptune. Maybe one does exist, but presently, the belief is that it does not.

PLUTO'S SATELLITE

What is the planet farthest from the Sun?

In 1999, Pluto will become the farthest planet from the Sun and will remain so for the next two centuries. In 1979, Neptune became the most distant planet, and it will hold that position between now and 1999. This can happen because the planets move in ellipses in which there is an aphelion (greatest distance from the Sun) and perihelion (least distance from the Sun). In 1979, Neptune was at aphelion, some 4.34 billion kilometers from the Sun, and Pluto was at perihelion, some 4.18 billion kilometers. Pluto was closer than Neptune.

Pluto is far out in the solar system, and it is small—only about 3000 kilometers in diameter, even smaller than the Moon. Because it is small and distant, Pluto is a dim object beyond the range of even most larger telescopes. It was discovered in 1930 by Clyde Tombaugh, who studied thousands of sky photographs that had been taken over a long period of time. Decades had been spent searching for an image that shifted slightly among background stars. In that year, Tombaugh was certain he had found the elusive planet.

Its existence had been suspected because neither Uranus nor

Neptune moved as predicted. It seemed that an object out there was altering their motions. That's why the search for Pluto began.

However, it turns out that Pluto does not have enough mass to cause the changes. There has to be another explanation. Some people suspect there is another planet out beyond Pluto.

How was Charon discovered?

In 1978, James Christy, an astronomer at the Naval Observatory in Flagstaff, Arizona noticed a slight bump on an image of Pluto. He found the same bump on other photos of the planet. He proposed that the bump was a satellite of Pluto which merged into photos of the planet. He named the satellite Charon.

In mythology, Pluto is god of the underworld. Charon is the boatman who ferries souls across the river Styx to appear before Pluto.

In relation to its planet, Charon is the largest of all the satellites. It is about 1200 kilometers across, making it almost half the size of Pluto. Its mass is about one-tenth that of Pluto. By comparison, the mass of the Moon is about one-eightieth that of Earth. Pluto and Charon are often thought of as a double-planet system.

The density of the two, the planet and its satellite, is very low, less than that of water. This means they are probably made mostly of ice—frozen gases. The implication is that Pluto, and perhaps Charon, at one time was a satellite of Neptune. Perhaps Pluto collided with Triton, or moved close to Triton, affecting it enough to cause Triton to move retrograde, the motion we observe today. At the same time, Pluto may have been deflected and thrown into its present orbit. Charon may have been carried along with Pluto.

A series of close encounters may have occurred way out there at the end of the solar system. Or is it the end? There may

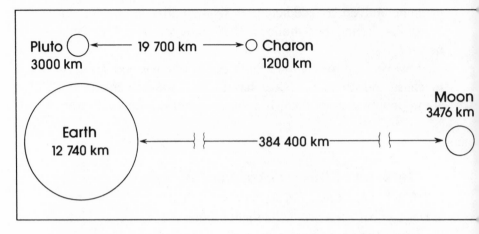

Compared to its planet, Charon is very large—nearly half the diameter of its planet. And it is located close to Pluto.

Comparatively, the Moon is one-fourth the size of Earth—and twenty times as far away from its planet.

be another planet beyond Pluto. We may be one of ten planets in the solar system, rather than the nine we have come to know. That's only one of many mysteries about the planets and their satellites that astronomers continue to explore.

CONVERSION TABLE

The metric system of measurement uses:
 meters for length
 grams for mass (weight at sea level)
 liters for volume

Some equivalent measurements:
 1 inch = 2.54 centimeters
 1 foot = 0.305 meters
 1 yard = 0.9144 meters
 1 mile = 1.609 kilometers
 1 pound = 0.454 kilograms
 1 quart = 0.946 liters
 1 centimeter = 0.3937 inches
 1 meter = 39.37 inches
 1 kilometer = 0.621 mile
 1 gram = 0.035 ounce
 1 kilogram = 2.20 pounds
 1 liter = 1.06 quarts

To convert U.S. customary units to metric, multiply by the factor indicated. For example:
 20 inches $\times$ 2.54 centimeters = 50.8 centimeters
 50 yards $\times$ 0.9144 meters = 45.7 meters

To convert metric units to U.S. customary units, divide by the factor. For example:
 500 kilometers $\div$ 1.609 = 310 miles
 35 kilograms $\div$ 2.20 = 11.36 pounds

FURTHER READING

Branley, Franklyn M. *Jupiter: King of the Gods, Giant of the Planets.* New York: Lodestar Books, 1981.

———. *The Moon: Earth's Natural Satellite.* New York: Thomas Y. Crowell, 1960.

———. *Saturn: The Spectacular Planet.* New York: Thomas Y. Crowell, 1983.

O'Leary, Brian, and Beatty, J. Kelly, editors. *The New Solar System.* Cambridge, Massachusetts: Sky Publishing Corp., 1981.

Smoluchowski, Roman. *The Solar System.* New York: Scientific American Books, 1983.

INDEX

Page numbers in *italics* refer to captions.

ABOUT THE AUTHOR

FRANKLYN M. BRANLEY is the popular author of over one hundred books for young people on astronomy and other sciences. His books include *Mysteries of the Universe* and *Mysteries of Outer Space* in the Mysteries of the Universe Series, *Halley: Comet 1986, Space Colony,* and *Jupiter.*

Dr. Branley is Astronomer Emeritus and former chairman of The American Museum–Hayden Planetarium. He and his wife live in Sag Harbor, New York.

ABOUT THE ILLUSTRATOR

SALLY J. BENSUSEN, illustrator of *Mysteries of the Universe, Mysteries of Outer Space,* and *Halley: Comet 1986,* has done work for the Smithsonian as well as for many astronomy magazines. She lives in Lanham, Maryland.

Branley, Franklyn Mansfield, 1915-
   Mysteries of the satellites /
Franklyn M. Branley ; diagrams by Sally
J. Bensusen. -- New York : Lodestar
Books, c1986.
   71 p. : ill. ; 24 cm. -- (Mysteries
of the universe series)
   Bibliography: p. 66.
   Includes index.
   Summary: Discusses the origins and
characteristics of the natural
satellites that travel around seven of
the nine planets.
   ISBN 0-525-67176-5
   1. Satellites--Juvenile literature.
I. Bensusen, Sally J. II. Title